Roaming Under the Blue Sky

LAN TIAN HOTEL

走进蓝天 蓝天酒店

蓝天酒店 编

華中科技大學出版社
http://www.hustp.com
中国·武汉

T
蓝天酒店
LANTIAN HOTEL

Lantian Hotel

This hotel is located on No. 203 of Yingxiongshan Rd., Jinan City, Shandong Province. The construction area is 13,000 square meters. This is a high level foreign hotel integrating accommodation, dieting, business, conference and other functions. It has superior geological location with convenient transportations. The column design of the exterior displays the magnificent momentum of the hotel. Generally speaking, European design represents luxury and elegance, interpreting some life attitude, some taste and some style.

LAN TIA

N HOTEL

Current architecture and the decorative theme of space design have progressed to a stage not necessarily in line with the style of a specific historical period, freely integrating styles of different periods. Both Chinese Neo-Classical style and European Neo-Classical style possess dual temperament which are classical and modern. Neo-Classical style applies much conciser and open-style approach to meet with people's longing for classical style, while, maintaining modern aesthetic effects. They are sometimes solemn and magnificent, sometimes fashionable and elegant, sometimes fresh and fluent. But all attain profound cultural connotations and aristocratic aesthetic tastes, which are the essence of Neo-Classical style. The design of Lantian Hotel salutes ancient glorious civilization with innovative ideas and concise approaches, starting from modern people's archaic affections and aesthetic interests and continuing classical space's elegance and charming appearance. With new techniques, new materials, new lights and new approaches, this hotel design reconstructs and reinterprets classical style, creating a hotel exclusive to this modern world.

Hotel architecture is an important business design field in the interior design industry. On one hand, hotel design needs to give the travelers some human cares just like home design. On the other hand, hotel design includes the elements of many public architecture designs. Lantian Hotel which is located on No. 203 of Yingxiongshan Rd. of Jinan City with total construction area of 13,000 square meters carries out perfect interpretation for this connotation. This is a high-standard foreign hotel integrating multiple functions, such as accommodation, dieting, commerce, conference, etc. The space seeking for orders and restrictions in classical art entrusts the architecture and space historical feel and human atmosphere through application of layout, lines and various soft decorations. While the hotel's exterior design presents modern architectural style. The column design of the exterior displays the uncommon momentum, perfect presentation of the integration of traditional elements and modern forms. With elegant design language and meticulous details, the guests' rooms demonstrate the graceful taste of the inhabitants. The classical style of the presidential suite is gorgeous and magnificent, in harmony with modern people's life texture. This hotel not only displays the magnificence of business hotel, but also possesses the romantic and warm atmosphere of holiday leisure hotel. The hotel space perfectly displays some optimistic life attitudes, some elegant cultural taste and some extreme space tones.

This hotel treats classical decorative style with new aesthetic viewpoints, combines archaic romantic affections with modern people's requirements for life with diversified thinking pattern and makes classical beauty transcend the time integrating elegance and modern fashion, displaying all lively aspects in modern life. This hotel restores classical temperament with modern design approach and concepts and creates pleasing high-quality space... Welcome to Lantian Hotel. Here ,you can fully experience the spectacular human temperament possessed by artistic aesthetic viewpoints and high-quality life of cultural tastes.

August 2013

Preface 序言

当下建筑、空间设计的装饰主题已经进化到无须与特定历史时期的风格相一致，可以自由地融合各个不同时期风格的特点。不管是中式新古典风格还是欧式新古典风格，它们都具备古典与现代的双重气质。新古典风格用更加简洁、开放的手法满足了人们对古典的怀念，同时，又拥有现代的审美效果。它们或庄重华丽、或时尚典雅、或清新流畅，但却都有着深厚的文化底蕴和贵族式的审美品位，而这正是新古典风格的精髓所在。蓝天酒店的设计以创新意识、简约手法向往昔辉煌的文明致敬，从现代人的复古情思和审美趣味出发，延续了古典空间的优雅高贵与迷人面貌，运用新技术、新材料、新灯光、新手法，对经典进行重新解构和演绎，创造出属于这个时代的“蓝天酒店”。

酒店建筑是室内设计中很重要的一个商业类型。一方面酒店设计需要如家居设计一样给旅人以人性化的关怀，另一方面酒店设计也包含很多公共建筑设计的元素。位于济南市英雄山路203号，建筑面积1.3万平方米，集住宿、餐饮、商务、会议于一体的高档涉外的蓝天酒店。对这种内涵进行了完美的诠释，空间中对于古典主义的艺术寻求条理和克制，通过布局、线条和各种软装的使用，赋予了建筑和空间历史感与人文气息。同时，酒店的外观设计展现出现代建筑风格，外立面的立柱设计彰显出酒店非凡的气势，是传统元素与现代形式融合的完美演绎。客房以优美的设计语言和考究的细节力求体现出居住者的高雅品位。总统套房的古典风格雍荣华丽令人迷恋，与现代人的生活质感不谋而合。蓝天酒店不仅具有商务酒店的大气奢华，还独具度假休闲酒店的温馨浪漫，使得一种积极的生活态度、一种高雅的文化品位和一种极致的空间格调得以恰到好处地表达。

蓝天酒店，用新的美学观点去审视古典主义的装饰风格，用多元化的思考方式，将怀古的浪漫情怀与现代人对生活的需求相结合，兼容华贵典雅与时尚现代，让古典的美丽穿越岁月，在现代生活中活色生香。用现代的设计手法和理念还原古典气质，打造宜人的高品质空间……到蓝天酒店来，可以让您充分地领略充满艺术化的美学观点和极具文化品位的高品质生活所应具有的独特人文气质。

2013年8月

Contents

目录

LAN TIAN HOTEL

The Exterior of Lantian Hotel

蓝天酒店外景

The Exterior of Hotel

酒店外景

济南蓝天酒店位于济南市英雄山路 203 号，建筑面积 1.3 万平方米，是一家集住宿、餐饮、商务、会议于一体的高档涉外酒店，坐拥济南市英雄山路核心商圈，交通便利，地理位置十分优越。酒店的外观采用欧式的设计风格，外立面的立柱设计彰显出酒店非凡的气势。欧式设计通常代表着奢华、优雅，它诠释着一种生活态度、一种品位和一种格调。蓝天酒店的外观设计展现出现代建筑风格，是传统与现代奢华融合的完美演绎。

This hotel is located on No. 203 of Yingxiongshan Rd., Jinan City, Shandong Province. The construction area is 13,000 square meters. This is a high level foreign hotel integrating accommodation, dieting, business, conference and other functions. It has superior geological location with convenient transportations. The hotel's exterior has European design style. The column design of the exterior displays the magnificent momentum of the hotel. Generally speaking, European design represents luxury and elegance, interpreting some life attitude, some taste and some style. The exterior design of Lantian Hotel presents modern architectural style, perfect representation of the integration of traditional and modern luxury style.

AN TIAN HOTE
蓝天

Exterior

入口处的建筑外观采用了丰富而又细腻的花草雕花，极富欧洲风情，以浅咖色和褐色为主体的建筑，通过小面积金色的点缀，呈现出充满质感的视觉体验。

The architectural exterior of the entrance applies abundant and refined flowers and plants carvings, full of European charms. Through ornaments of small area gold colors, the building with light coffee color and brown color as the main colors presents visual experiences full of texture.

LAN TIAN HOTEL

蓝天酒店的入口处设计简洁有力，体现了东西方建筑文化的交融。外敞式的柱子在视觉上有着强烈的形式感，彰显着酒店开放的意识形态，而精致的铁艺栏杆则传达出酒店细致、典雅的一面。周边环境静谧、优雅，堪称“闹中取静”。蓝天酒店完美的结合了品牌的精致奢华与身处都市的活力。

The entrance design of the hotel is concise but powerful, displaying the integration of eastern and western architectural culture. The column of open style has intensive formal sense visually, highlighting the open ideology of the hotel. And the delicate iron balustrade conveys the exquisite and elegant aspect of the hotel.

LAN TIAN HOTEL

Lantian Hotel Lobby

蓝天酒店酒店大堂

T. ARMSTRONG & BRO
MANCHESTER
北京
BEI JING
东京
TOKYO

Hotel Lobby 酒店大堂

走入酒店的大堂，就被大堂宽敞的空间、恢弘的气势所吸引。设计秉承了低调的奢华与传统的中国风和谐统一的宗旨和理念。设计师将传统的中式廊架引入室内作为休息区域，让空间充满传统韵味。在这里，每一位客人都将感受到世外桃源的清静，以及对休闲的酒店居住环境感到由衷的向往。

Upon entering the hotel lobby, you would be fascinated by the lobby's extensive space and magnificent atmosphere. The design is in accordance with the harmonious integration of low-key luxury and traditional Chinese style, and takes it as the tenet and concept. The designer introduces traditional Chinese gallery as the interior rest area, making the space full of traditional charms. Here, every guest can sense the purity and peace of the idyllic land, thus sincerely longing for the leisurely hotel residential environment.

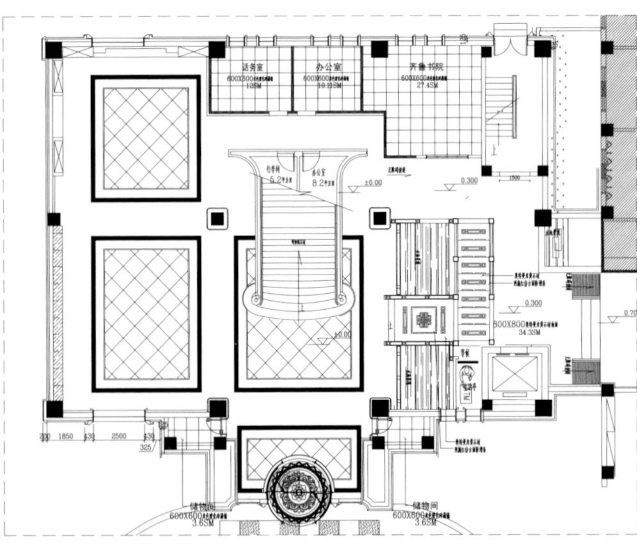
话务室
办公室
齐鲁书院
行李间
5.2
办公室
8.2
±0.00
0.300
1500
0.300
0.70
800X800
34.3SM
300
1850
430
2500
430
325
储物间
600X600
3.6SM
储物间
600X600
3.6SM
大堂铺装图

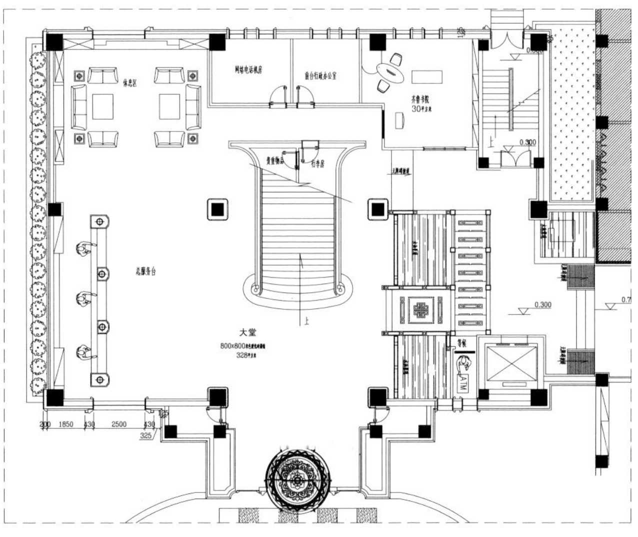
休息区
网络电话机房
前台行政办公室
齐鲁书院
30平方米
贵重物品
行李房
总服务台
大堂
328平方米
等候
ATM
0.300
200
1850
430
2500
430
325

大堂平面布置图

Lobby Lounge

大堂休息区

深色的皮革家具配上奢华的吊灯，再以鲜艳的花朵作为点缀，为大堂的休息区营造出舒适的氛围。为了突出蓝天酒店与众不同的特色，在一旁的展示柜上摆放着逼真的飞机模型。一个个栩栩如生的飞机模型不仅代表着时代的进步，也代表着人类对无边无际蓝天的向往之情。

The combination of dark color leather furniture and luxurious droplights, together with the ornaments of bright-colored flowers, creates comfortable atmosphere for the lobby lounge. In order to highlight the spectacular characteristics of the hotel, there are some vivid space shuttle models in the showcase on the side. These true-to-life models not only manifests the progress of the time, but also represents people's longing for the immense blue sky.

Archaic Charm Hall 古韵堂

抿一口清茶，赏一幅字画，悠闲的生活态度在此表露无遗。借着朦胧的灯光，古韵堂古朴的牌匾映入人们的眼帘。透着古色古香的隔扇，人们的思绪仿佛能穿越古今，伴着历史缓缓走过的年轮，回到让人安心的年代。

Leisurely life attitude is displayed to the utmost while enjoying a cup of tea and a piece of calligraphy and painting. Within the dim light, the archaic plaque of the hall leaps into the eyes of people. Through the screen of antique flavor, people's thoughts seem to traverse ancient and present time. With the wheels of history slowly passing by, people seem to go back to the peaceful past.

古韵堂

LAN TIAN HOTEL

The Public Space of Lantian Hotel

蓝天酒店公共空间

Public Space

公共空间

作为放松、娱乐之用的公共空间，是酒店中不可或缺的一部分。蓝天酒店的公共空间宽敞明亮，过厅中淡雅的色调、复古的装饰使得空间光亮而绚烂，为到来的客人带来了愉悦的心情。公共空间可供交际与休闲两用，您可以在这找到感兴趣的一切，艺术品和字画琳琅满目。与三五好友品品茶艺、在美酒汇中畅谈人生，您可以在这里度过惬意的时光。

The public space for relaxation and entertainment is an indispensable part of the hotel. The public space in the hotel is expansive and bright. The elegant color tone and archaic decorations of the gallery makes the space appear bright and magnificent, creating some pleasant feelings for the coming guests. The public space has both communication and leisure functions. You can find everything here you are interested in, artistic objects, calligraphy, paintings, etc. You can invite several friends to speak glowingly of life while enjoying tea or good wine. You can have a wonderful time here.

北京画坛
北京画坛
FACE LOOKING
SIDE HIDDEN

Good Wine Collection

美酒汇

品酒，作为高档的生活体验之一，越来越受到人们的欢迎。美酒汇中不仅提供各式各样的美酒，更是为品酒的客人打造出浪漫、私密的高雅环境。设计师注重以细节打动人，在这里，客人不但可以享受到让香醇的芬芳留在舌尖，更能享受到环境所带来的舒适感。

Wine-tasting, as one of high-level life experiences, has become more and more popular among people. Here, there are not only all kinds of good wines, but also that it creates a romantic and private elegant environment for the guests. The designer focuses on moving people with details. Here, the guests not only can enjoy the fragrant flavors, but also enjoy the comforts of the environment.

Public Rest Room

公共卫生间

卫生间延续了其他空间的风格特征，选用深色调的木质家具作为主要装饰。在正对着洗手台处搭配上色彩浓厚的花束，显示出油画般的质地，凸显出空间的雅致。洗手台旁采用一整面墙的镜面设计，在视觉上扩大了空间感。卫生间的整体设计以高雅而不经意的内部装饰呈现出恰如其分的精致。

The rest room continues the features of other spaces, selecting dark color wooden furniture as the main decorations. There is a bouquet of bright flowers on the opposite of the wash basin, highlighting the elegance of the space with oil-painting-like texture. The wash basin applies whole wall mirror design, visually expanding the space. The whole design of the rest room presents the appropriate delicacy with elegant but casual interior decorations.

LAN TIAN HOTEL

Catering Area

餐饮区

Buffet Area

自 助 餐 区

自助餐区一向是客人们青睐有加的地方。美食不仅可以补充能量，还能舒缓身心，让人心情愉悦。自助餐区也是适合人们沟通交流的地方，宽敞的空间以利于人们活动为宜，桌椅的摆放尽量靠近窗边，让客人在品尝美食的同时，还能欣赏到窗外的美景。

Buffet area has always been a place gaining popularity among the guests. Gourmet can not only supplement energy, but also release physical and psychological pressure, making people feel happy and pleased. Buffet area is also a place for people to communicate with each other. The expansive space is appropriate for people to carry out activities. The desks and chairs are placed near the window. Thus, while enjoying the gourmet, people can feast also on the nice sceneries outside.

蓝天酒店的自助餐区设计得典雅大气，运用素雅的色调强调用餐区的整洁及用餐氛围的平和。桌椅的摆放分为普通四人座和适合聚餐的十人座，满足客人的多种需要。空间一旁有整排的落地窗户，既能够给室内带来良好的采光，也方便窗内外空间的相互沟通。自助餐台上，中西方的美食应有尽有，不少客人都难以抵挡这里上演的美食诱惑。

The buffet area design of the hotel is elegant and magnificent, with elegant color tone manifesting the clarity of the dining space and peaceful dining atmosphere. As for the layout of tables and chairs, there are tables for four and party tables for ten, meeting the different requirements of the guests. There is a whole row of French windows on the side of the space, bringing in fine daylighting for the interior space, while making it convenient for the mutual communication between the interior and the exterior. On the buffet table, you can find all kinds of eastern and western gourmet. Many guests can not resist the temptation of gourmet here.

1964年7月23日,毛泽东主席
英雄岳振华时，握着他的手说：
打的好，打的好啊！

Private Room

包 间

走进蓝天酒店的包间，就觉得处处流露着典雅和气派。包间采用简欧的风格，注重加强细节的特点来装饰整体环境，包间空间宽敞、环境优美。蓝天酒店给客人提供如家一般舒适的包间，让客人在此尽情体验会客、用餐的便利。

Upon entering the private room of the hotel, you would find that elegant and magnificent atmosphere is everywhere. The private room is concise European style, emphasizing on the features of detailed parts, thus decorating the whole environment. The private room has expansive space and nice environment. This hotel provides the guests with private rooms where they can feel like home, making them enjoy to hearts' contents convenience of meeting with friends and dining.

包间 – 蓝之本

蓝之本包间运用纯净色彩的大理石立柱来装饰空间，立柱经过精心雕琢，展现出欧式的古典与恬静，用色简单，却别有一番风情。墙壁上的装饰挂画也与主题相关，在蓝天中自由飞翔的飞机，承载着对蓝天最本质的梦想。

This private room applies marble column of pure colors to decorate the space. After exquisite carving, the column manifests classical and peaceful European style. Although the color is simple, it presents some peculiar charms.

包间－蓝之翼

蓝之翼包间有着深沉的色调，白色餐布配上深色的座椅对比强烈，更加凸显了空间的纯净。金属质感的吊灯璀璨夺目，是空间的焦点所在，分外引人注目。

This private room has profound color tone. The intensive contrast between white table cloth and dark chairs highlight the purity of the space. The chandelier with metal texture is the focus of the space, quite enchanting.

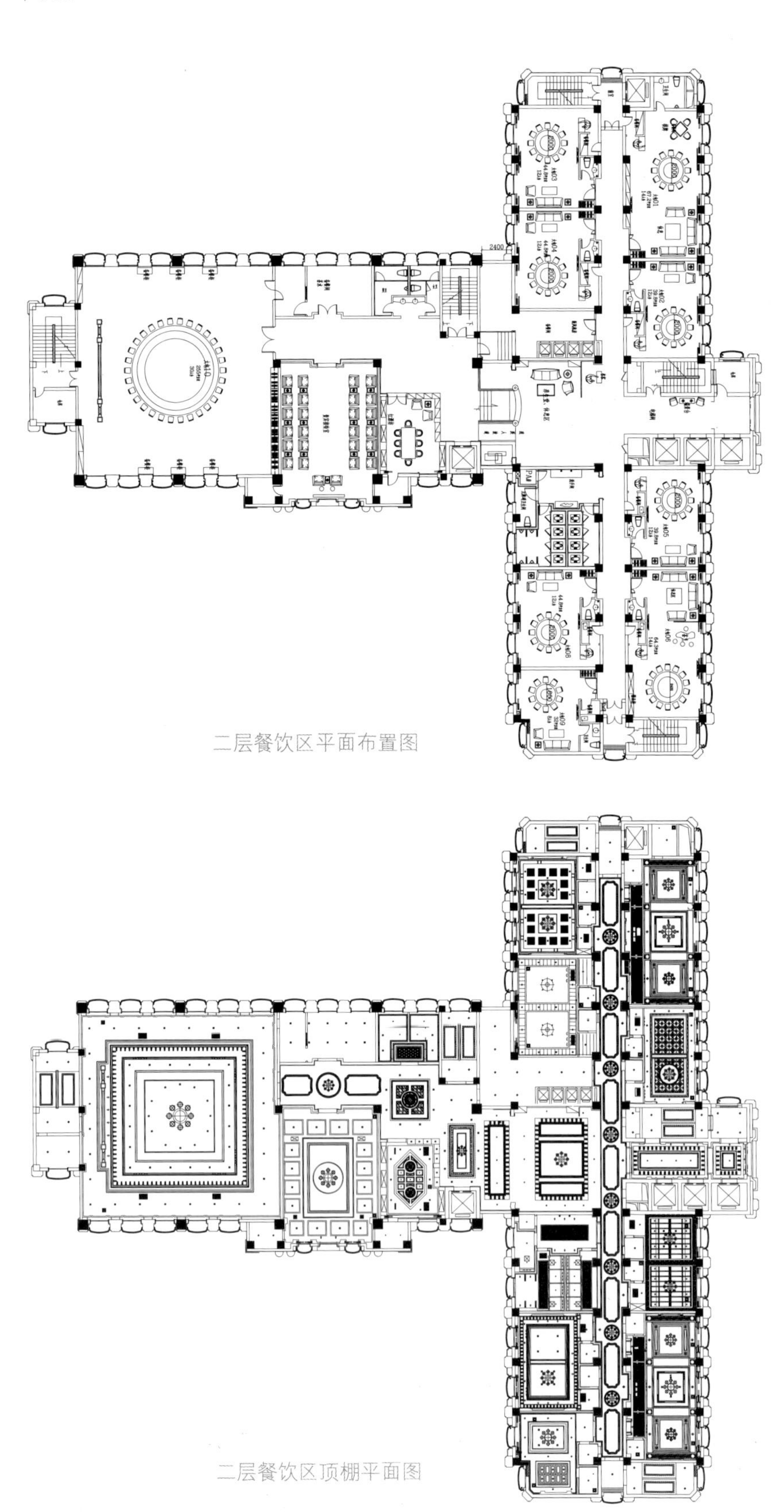

二层餐饮区平面布置图

二层餐饮区顶棚平面图

歼-5飞机 Jian-5 FeiJi

包间 – 蓝之歌

包间－蓝之蕾

包间 – 蓝之梦

蓝之梦是蓝天酒店中最大的一个包间，类似宫廷宴会式的餐桌形式，彰显出此空间的庄重感。间隔着落地窗的墙面镶嵌着金色的装饰构件，整体环境低调而又奢华。

This is the largest private room inside the hotel, quite similar to the table form of palace banquet, displaying the solemn atmosphere of the space. The wall of the French window is inlaid with gold decorative objects, demonstrating the low key but luxurious whole environment.

包间－蓝之夜

空间采用明黄色的壁纸与褐色的木质边框相搭配，展现出明快的色彩氛围。室内还摆放着色彩绚烂的花束，为室内空间带来自然的气息，也为用餐者提供了一个愉悦的空间环境。

The space applies the combination of bright yellow wallpaper and brown wooden frame, displaying brisk color atmosphere. There are bright-colored bouquets inside the space, creating some natural atmosphere for the space, while at the same time providing a pleasing dining environment.

LAN TIAN HOTEL

Meeting Area

会议区

VIP Reception Hall

贵宾接待厅

贵宾接待厅为宾客提供了一个完美的私人会谈或是商务会晤的场所。透过一旁的落地玻璃窗，客人可以饱览花园美景，尽享轻松惬意。空间为宾客营造了一个放松心情、宁静私密的环境。

The VIP Reception Hall provides a perfect place for private meeting and business conference. Through the side French window, the guests can have a great view of the garden landscape, while having leisure and ease. The space creates a relaxing and tranquil space for guests.

贵宾接待厅

浅色的色调让空间具有一种安定、平和之感。空间较为素雅，没有过多的颜色装饰来分散人的视线，但又在顶棚处延续了其他空间的金色进行了适当的装饰，使这里适合工作洽谈。

The light color tone creates some tranquil and gentle atmosphere inside the space. There is no excessive color decorations to distract people's eyesights, but that the designer gets some appropriate decoration on the ceiling to continue the gold color of other spaces, which makes the space perfect for negotiation.

1958年2月13日，毛泽东
主席在沈阳飞机制造厂视察歼-5
斗机生产线

天宫梦圆

中国女飞志凌云

藍 天
LANTIAN HOTEL

Lantian Hall

蓝天厅

蓝天厅以现代风格装饰，入口处的实木门彰显出蓝天厅的气派，室内空间宽敞、色调淡雅，可满足各类活动需要。是举行重要商务会议和庆祝宴会的理想场所。

This hall is decorated of modern style. The solid wood door at the entrance manifests the magnificence of the hall. It has expansive interior space and elegant color tone, which allows it to meet with various kinds of activities. It is a perfect location to hold important business conferences and celebration banquets.

Lan Tian Hotel
蓝天酒店

Conference Mode

蓝天厅／会议模式

蓝天厅的会议模式配备了先进的设施。礼堂式的大厅提供高科技的影音设备、用于多媒体展示的投影仪和大屏幕、电脑展示切换功能接口及大屏幕等离子电视。会议厅宽敞别致，主席台及背景精心设计，坐在整齐摆放的座椅上能欣赏到主席台上的精彩瞬间。

The hall's conference mode is equipped with advanced equipments. The auditorium-like hall provides high-tech audiovisual equipment, projector and large screen for multimedia presentation, functional interface for computer presentation switch and large screen PDP. The conference hall is expansive and unique. The rostrum and background are well designed. The seats neatly placed make it possible for people to have a full view of the rostrum.

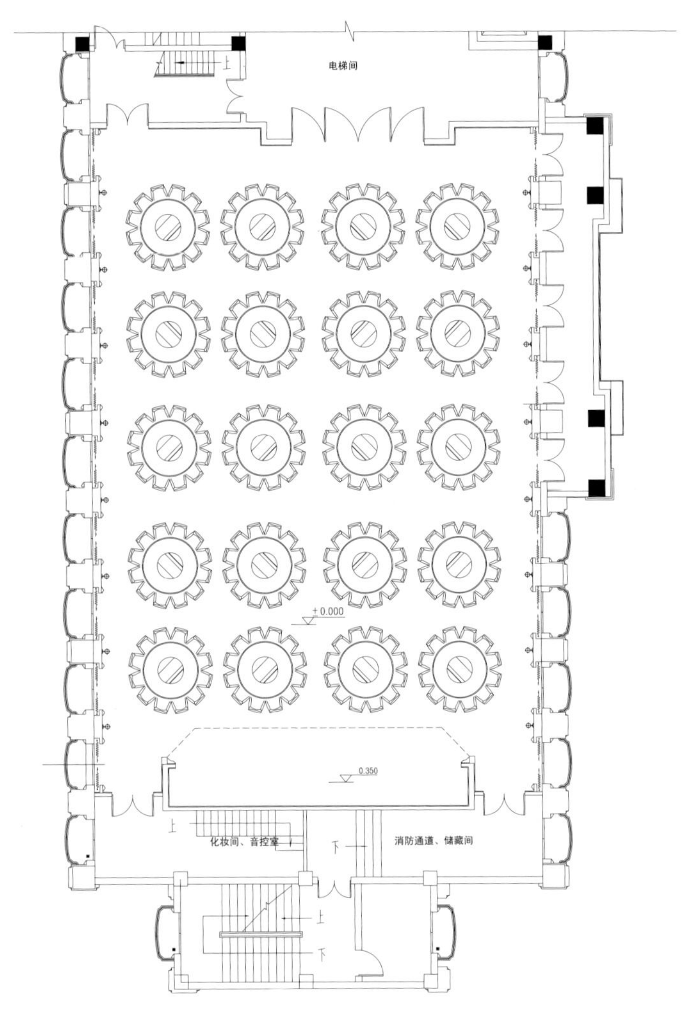

蓝天厅平面布置图

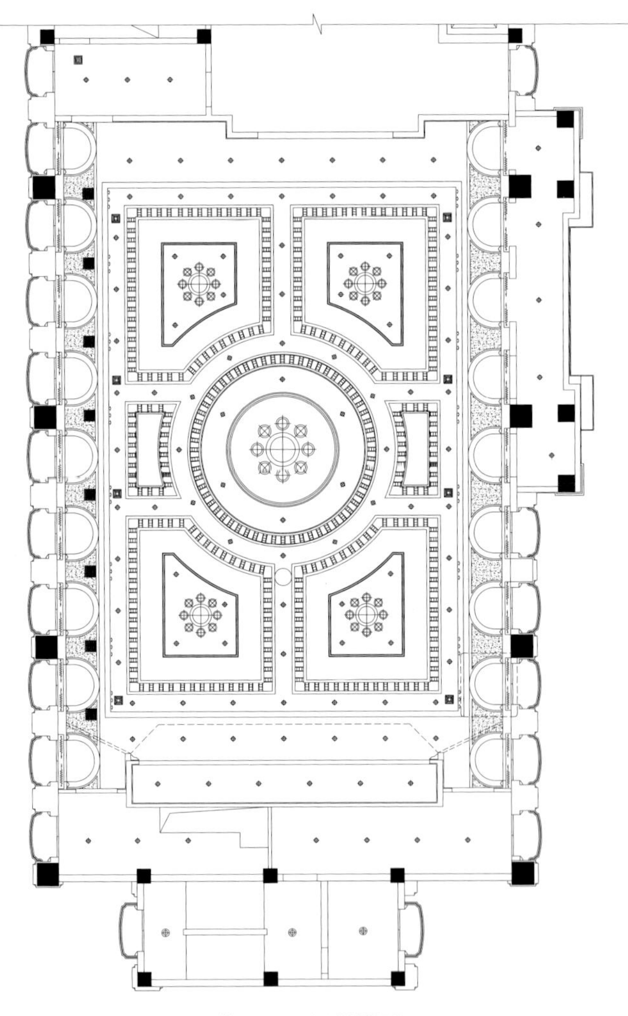

蓝天厅顶棚平面图

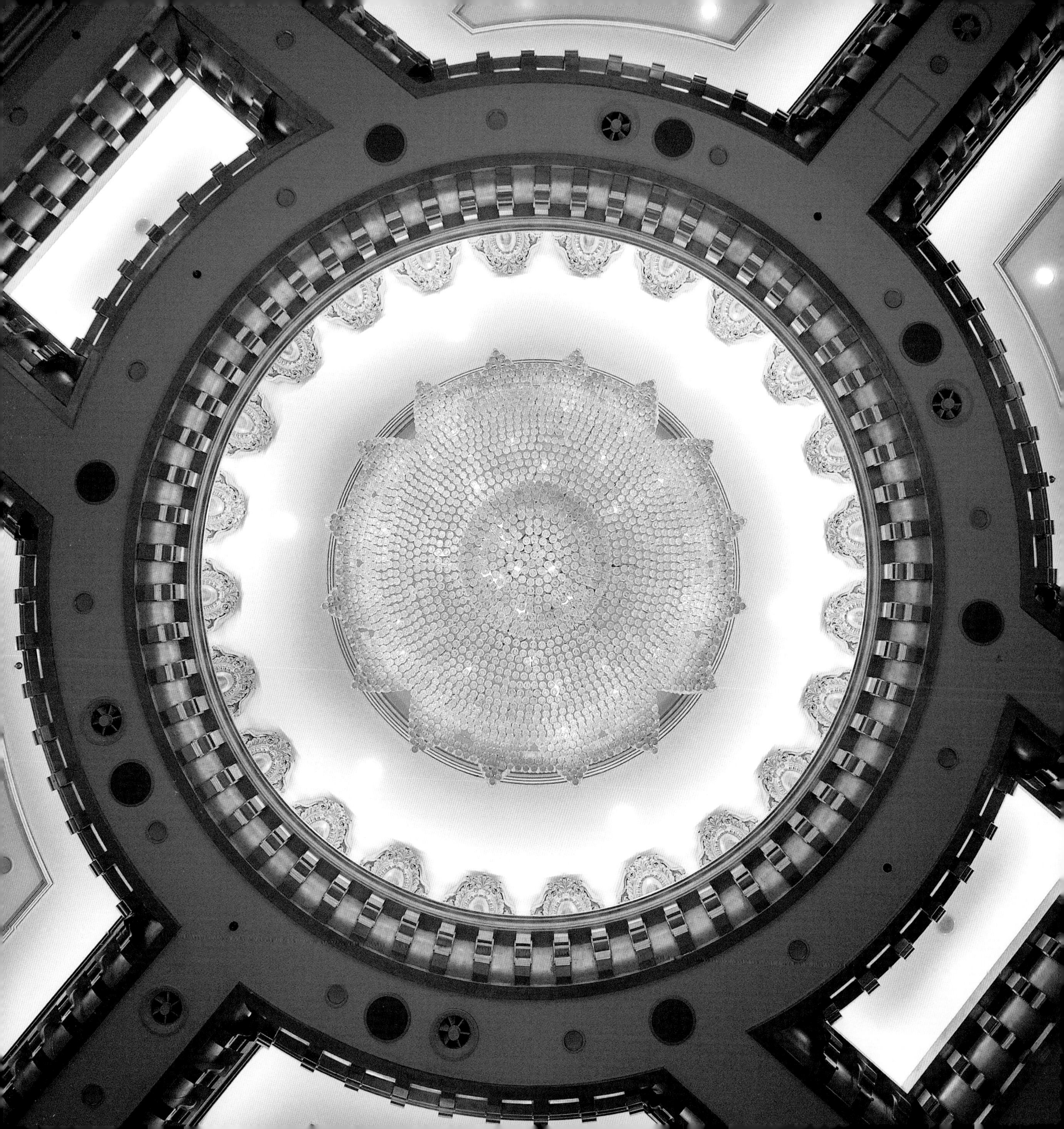

Hotel

蓝天厅立面图一

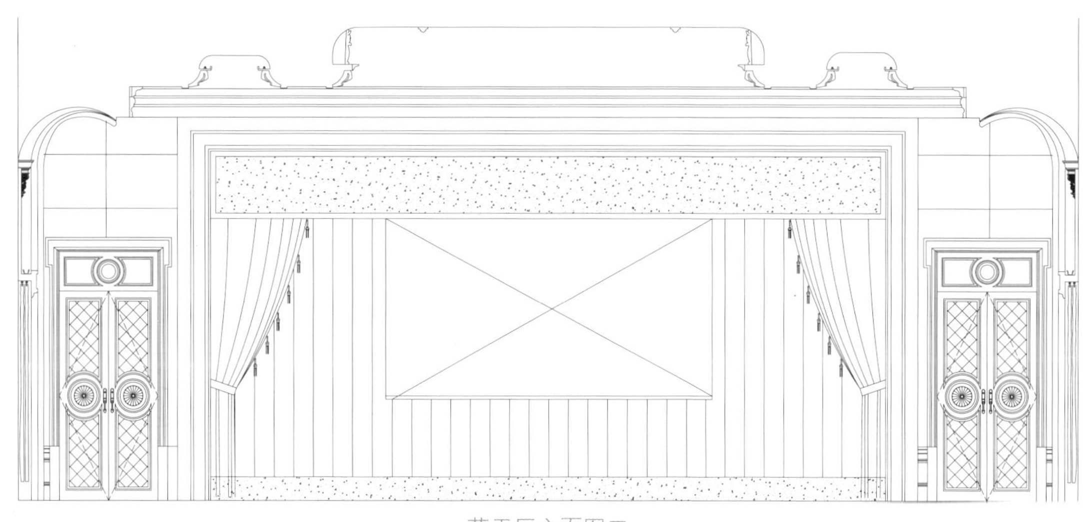

蓝天厅立面图二

蓝天厅立面图三

LanTian Hotel
蓝天酒店

Banquet Mode

蓝天厅／宴会模式

蓝天厅的空间宽敞，可以很好地满足宴会的需要。室内空间的两侧均有落地窗户，有充沛的天然采光。 设计赋予酒店空间的精神自中央的水晶吊灯循序展开，古典雅致的装饰纹样在不断重复中赋予空间优雅而又庄重的感觉，顶棚的设计成为贯穿空间的精彩的一笔。

The hall has broad space, well satisfying the demands of banquet. There are French windows on both sides of the interior space, introducing sufficient natural lighting. The design spirit of the hotel space extends from the central crystal chandeliers. The continuous repetition of classical and elegant decorative pattern entrusts the space with elegant but solemn sensations. The ceiling design becomes the highlight connecting the whole space.

出口
EXIT

EXIT

Glenfiddich
12
XO

Coffee Bar

咖 啡 吧

咖啡吧是一个较为别致的空间。深色的木质家具弥漫出慵懒的味道。厚实、舒适的沙发座搭配室内微弱的灯光环境，让人可以在此静静享受属于自己的时光。步入咖啡厅，映入眼帘的是比较复古感的吧台，造型古典、稳重，木质暗哑的光泽与玻璃杯透彻质感的搭配，使得一切看起来都具有和谐感。吧台背柜，采用稳重的重色系木材，给人以厚重的感觉。几何形式感强烈的顶棚造型有意识地区分开吧台和卡座的区域，吧台的暗色系与卡座的明色系在空间中形成鲜明对比，但又不失统一。

The coffee bar is quite a unique space. Dark color wooden furniture sends out some lazy sensations. The combination of thick, solid and comfortable sofa and dim interior lighting environment makes it possible for people to quietly enjoy the time of their own. Upon entering the coffee hall, what leaps into the eyes first is the archaic style bar counter, with classical and sedate design. The dull wooden luster combines with the transparent texture of glass cup, making everything appear harmonious. The back cabinet of the bar counter applies sedate dark color wood, which produces some dignified sensations. The ceiling format of intensive geometrical style consciously divides the bar counter from the seats. The dark color of the bar counter and the bright color of the seats form some brisk contrast, while maintaining integrity.

EXIT

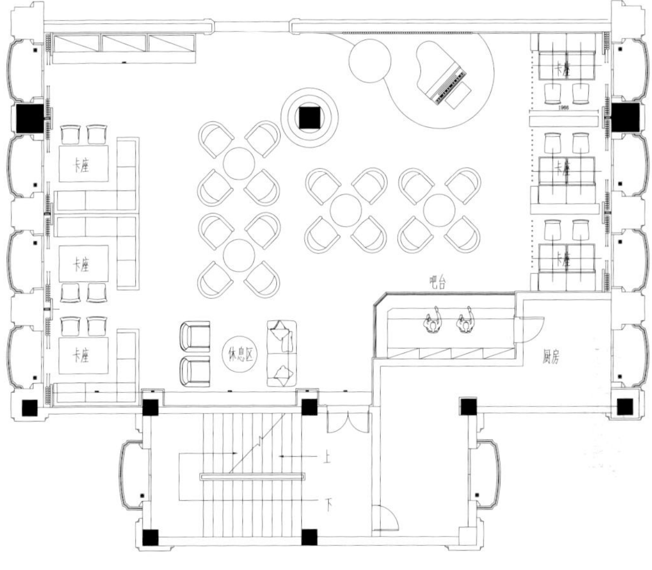

咖啡吧平面布置图

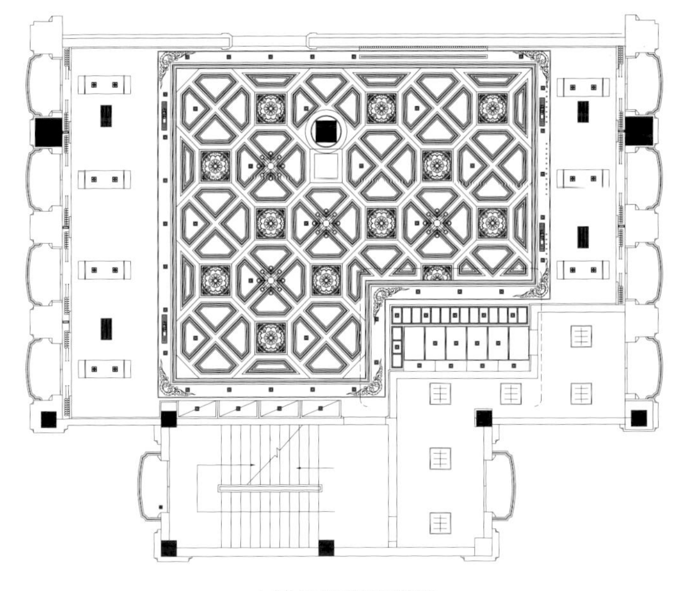

咖啡吧顶棚平面图

LAN TIAN HOTEL

Guest Room

客房

Guest Room

客房标间

客房是客人使用最长时间，也是最能体现酒店质感的空间。蓝天酒店的客房都经过精心设计，现代装饰风格中融汇令人不胜枚举的精巧细节，更有超大的落地窗供宾客欣赏美景。 每一间都风格典雅、配置奢华，为客人们提供了远离都市喧闹的最佳之所。

Guest room is a space where the guests would spend the longest time and which can represent the hotel's texture. All the hotel's guest rooms are designed meticulously. Modern decorative style integrates limitless delicate details, with large French windows which allow the guests to enjoy the fine view outside. Every guest room is elegant in the style and luxurious in the equipments, providing a perfect location for the guests to get far away from the metropolitan hustle and bustle.

Suite 509

套房 509

套房 509 采用中式风格，室内设计舒适、典雅，内饰采用木质品，渲染出浓厚的现代亚洲风情。米黄色的墙面与木色相搭配，突出平静、祥和的空间氛围。透过精巧的木质隔断，可以看到书房的景物，空间隔而不断。

Suite 509 applies traditional Chinese style, with comfortable and elegant interior design. The ornaments are wood objects to display intensive modern Asian amorous feelings. Beige wall combines with the wood color to highlight the peaceful and quiet space atmosphere. Through the delicate wood partition, one can see the views inside the study. Although there is partition, the space is still a whole.

书房的布置十分简洁，一桌一椅均是简单的木质样式，摒弃繁杂的设计，让空间显得开阔。书房拥有良好的采光，无论在清晨还是黄昏，客人都可透过落地窗欣赏窗外的美景。

The layout of the study is quite concise. The table and the chair are all simple wood format. The space appears quite spacious with no complicated design. The study has fine lighting. Be it dawn or dusk, the guests can always enjoy the nice view outside the window through the French window.

卧室的设计洋溢着自然的芬芳，用怡人的植物挂画来装饰，让客人在此仿佛回归了自然的怀抱。深木色的床给人一种安定的气息，简单而舒缓的色调容易让人放松心情，营造出一个良好的睡眠环境。

The bedroom is overflowing with natural fragrance, decorated with pleasing plant painting hung on the wall, which makes the guests seem to recover the original simplicity. The bed of dark wood color tone creates some quieting atmosphere for the guests, and the simple but solacing color tone can easily get people relaxed, though creating a perfect sleeping environment.

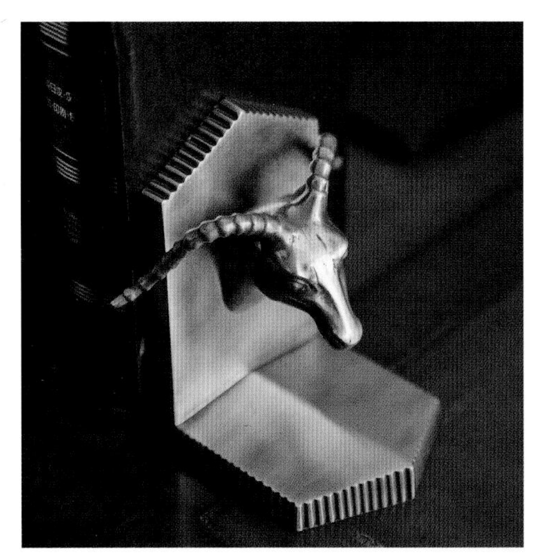

Suite 518

套房 518

套房别具气势的落地长窗设计，让客人得以饱览城市如画的风光或酒店漂亮的庭院景致。此间客房的设施十分现代，可以满足日常所需的一切。客房的设计温馨如家，整体铺装的地毯为室内减少了噪声，暖色调的室内氛围让空间与人更加亲近。

The magnificent long French window design of the suite allows all the guests to have a good view of the urban picturesque landscape and the hotel's nice courtyard sceneries. The modern equipments of the guest room can meet all daily requirements. The design is quite warm and sweet, the carpet can lessen the interior noise, and the warm color interior atmosphere shortens the distance between space and the guests.

卧室与客厅风格一致，延续了时尚、高贵的简约风格。豪华的大床配置了丝绸质地的抱枕，繁复的花纹吸引着人们的眼球，恢宏大气的设计尽显尊贵。墙壁上淡淡的纹样与整体空间互为呼应，显现出一种低调的品位。

The style of the bedroom and the living room is consistent with eath other, continuing the fashionable and noble concise style. The luxurious bed is equipped with silk pillow, with enchanting complicated pattern. The magnificent design displays nobility to the full. The light pattern on the wall corresponds with the whole space, presenting some low-key taste.

Suite 609

套房 609

一走进这里，墙壁上几张黑白照片的装饰将人的思绪带入从前，空间中弥漫着一种复古的格调。客厅里装饰着白色的木门，低矮的沙发舒适而柔软，落地窗边洁白的帘幔随风摇曳，优雅而静谧。套房有着视野开阔的优点，推窗眺望，窗外景致尽收眼底。

Upon entering the room, your thoughts would be driven back to the past by the several black-and-white photographs on the wall. The space presents some archaic tone. The living room has a white wood door. The soft low sofa is quite cozy. The pure white window curtain of the French window sways in the wind, quite elegant and tranquil. The suite has the advantage of broad views. Opening the window, and all the sceneries would leap into the eyes.

112

SPIRIT OF ST. LOUIS

Guest Room

Suite 618

套房 618

套房 618 采用温暖的黄色系装潢，展现出美式风格的特征。设计师运用丰富的材质及地道的色调搭配，创造出温馨亲切的豪华酒店环境。 现代装饰风格中蕴含着精巧的设计细节，透过落地窗，客人可饱览窗外美景。

This suite applies warm yellow color decorations, presenting American style features. The designer applies the combination of rich materials and local color tones, to create warm and kind luxury hotel environment. The modern decorative style integrates limitless delicate details, with large French window which allows the guests to enjoy the views outside.

Guest Room

Suite 709

套房 709

舒适的房间是疲惫游客的休憩之地，该套房设计风格典雅、配置奢华，为游客们提供了远离都市喧闹的最佳之所。所有空间都精心设计，绿植在室内随处可见。喜爱花鸟虫鱼的客人们应该也会喜欢这个充满中式田园风情的小屋，在这个恬静的空间中，游客可以放松身心，静静思索旅行的目的。

Cozy room is a rest space for exhausted travelers. With elegant style and luxurious equipments, the suite is a perfect place for the guests to get far away from the noisy urban life. Every space has meticulous design, and you can find green plants everywhere inside the space. The guests who are fond of flowers, birds, insects and fish would fall in love with this cabin filled with traditional Chinese pastoral style. In this quiet space, the travelers can get fully relaxed and ponder on the purpose of this journey.

Suite 718

套房 718

套房 718 的室内布置独特：优雅的金属吊灯、定制的家具、丰富的面料和油画般色彩的花束，尽显奢华。其中，宽敞的尊贵豪华客厅以宜人的明黄色为基调，在这里，客人可以沐浴在从超大玻璃窗射入的自然光线之中，品味着设计将奢华与优雅完美地融为一体。

The interior layout of this suite is quite spectacular. All are luxurious, including elegant metal droplights, custom-made furniture, rich fabrics and painting-like bouquet. Here, the guests can get bathed in the natural lights coming through the large glass window, tasting the perfect integration of luxury and elegance of the design.

LG

优雅的室内装饰像是一幅风景画，而卧房正是这幅风景画的焦点。卧房设施完善，有家的舒适感。新设计的整体环境被赋予了温暖的色调，并配有舒适的现代家具和装饰物，典雅而平和。

Elegant interior decoration is like a landscape painting, and the bedroom is the focus of this painting. The bedroom has complete equipments, which makes people feel like home. The whole environment of the new design is entrusted with warm color tone. Accompanied with comfortable modern furniture and decorations, all appear elegant and peaceful.

江泽民主席接见试飞员

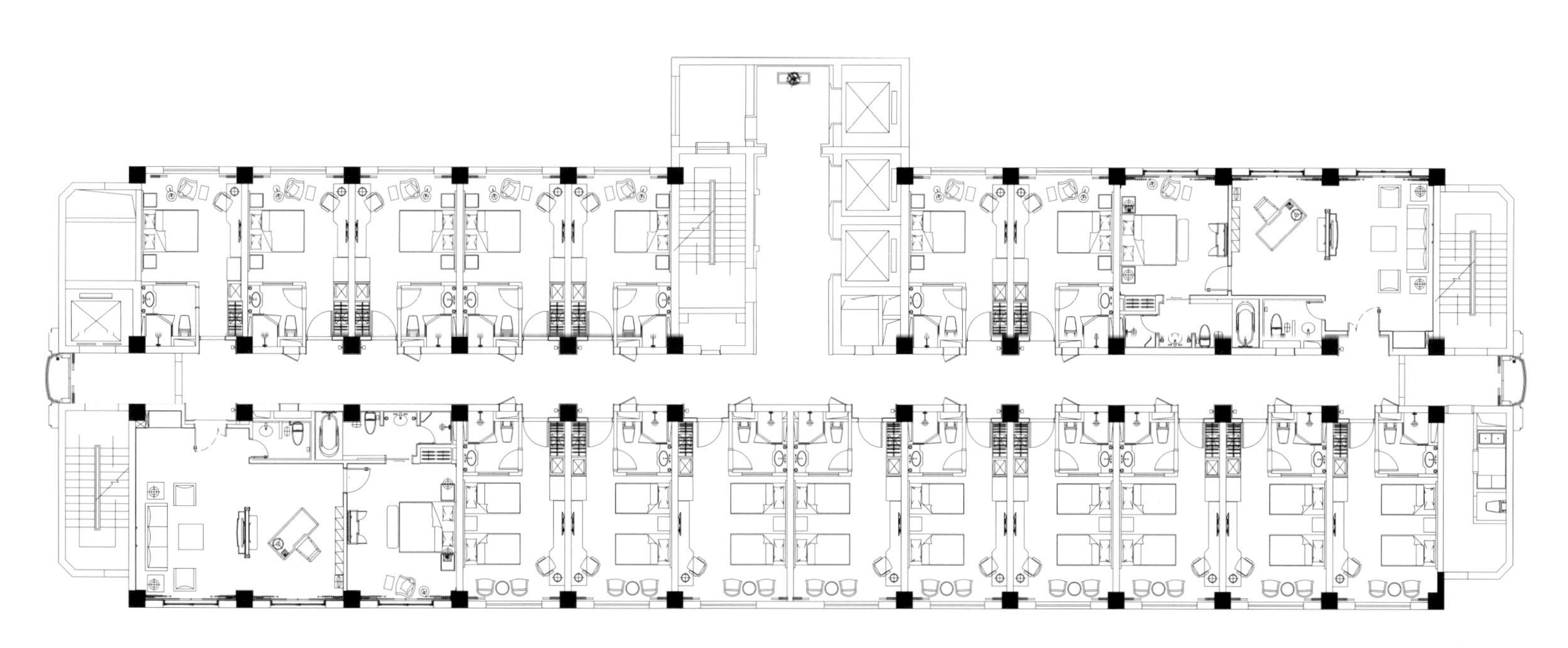

八层客房平面布置图

Suite 809

套房 809

新装修的套房色彩搭配巧妙，条纹状的壁纸搭配抽象画中的圆形，展现出一种天真的童趣。舒适的沙发为客人提供了舒适的休闲环境。套房中运用了多种材质进行装饰，辅以清新的绿植、透亮的玻璃饰品，加上轻快的灯光，使套房更显高贵。

The new decorated suite has ingenious color collocation. The combination of stripe-like wallpaper and the circle in the abstract painting displays some naive child interests. Comfortable sofa provides the guests with cozy leisure environment. The suite applies all kinds of materials for decoration. Accompanied with fresh plants, dazzling glass objects and brisk lights, the suite appears nobler.

Suite 818

套房 818

碎花图案的沙发透露出宁静的气韵、白瓷的灯具散发着温润的光芒，配上令人舒服的淡蓝色，在这些元素里深层挖掘便有了优雅精致的房间轮廓。颜色、材料和装饰的相互搭配，让不同空间能够高度融合在一起，整个空间充满着洁净、舒适的感觉。

Sofa of floral graphics send out tranquil charms. White porcelain lights emit warm lights. Accompanied with pleasing light blue color, all these elements manifest some elegant and delicate room silhouette. The mutual collocation of colors, materials and ornaments makes different spaces become a highly integrated whole. The whole space is full of pure and comforting sensations.

PARIS

窗外慵懒的阳光洒入进来，让旅途的生活也可以像“家”一样温暖。
饱含着理性情怀的蓝色更是让这种温暖发挥到极致。

The lazy sunshine from outside the window spreads inside, making the travel life as warm as home. The blue color of rational sensations makes this warmth display to the extremes.

Suite 909

套房 909

在材质的选择上，冷暖色调的材质在空间里丰富变换，让空间更加和谐丰满。

开阔的视野和典雅的观感让舒适、温馨弥漫每个角落。简约、柔和的灯光使客人沐浴在惬意的氛围中。

As for the selection of materials, the cold and warm color tone produces rich variations inside the space, making the space much more harmonious and complete. Broad views and elegant sensations make comfort and warmth pervade every corner. Concise and soft lights allow the guests to get submerged in the pleasant atmosphere.

Guest Room

多元的空间融入了亚洲风情的元素，硬装的简约设计为多元的软装素材提供了和谐的背景。

Multiple spaces are integrated with elements of Asian amorous sensations. Hard concise design provides a harmonious background for the multiple soft decoration materials.

Suite 918

套房 918

以简约的手法对空间进行装饰，将文化的底蕴表现出来，将奢华大气的氛围带入空间，为客人提供最优生活品质的生活享受。室内色彩以素雅大气的色调为主，辅以大地色系将庄重典雅的气势呈现出来，在通透的灯饰、金银铜器的点缀下，流露出岁月积淀的华美和温情。

The designer decorated the space with concise approaches, bringing out the cultural connotations and bringing luxurious and magnificent atmosphere inside the space, offering guests life enjoyment of superb life quality. The interior colors focus on elegant and grand color tones, accompanied with earth tone which brings out solemn and graceful momentum. With the ornaments of transparent lights, gold, silver and copper wares, the space is overflowing with the splendor and sweetness of the long past time.

Suite 1009

套房 1009

套房的整体空间均以浅淡的色彩来表现，营造出宛如湖面的婉约、雅致。加之明亮的灯光处理，更是将空间映衬得十分舒适，整体散发出一种真实的气息，带给人清爽、飘逸、灵动的感觉。实木质感的框架与家具则充分流露沉稳的气质，将古典的风韵尽显无遗，完善的空间氛围为客人打造出完美的生活。

The whole space of the suite is represented with light colors, creating some delicacy just like lake surface. Together with the bright lights, the space appears quite cozy, making the space send out some genuine senses and bringing out brisk, breezy and dynamic feelings. Solid wood frameworks and furniture fully send out some sedate temperament, completely manifesting the classical charms. The complete space atmosphere creates a perfect life for the guests.

卧房以木质为主，辅以布艺、墙纸等极具柔和气质的材料，给整个空间增添了不少生活情趣。空间格局简单而清晰，简单而利索的线条为空间平添几分立体感，加上灯光的晕染，更显得富有层次。空间凸显沉稳大方，再辅以精心搭配的软装细节，细腻而富有生活情调。

The bedroom focuses on wood materials. Accompanied with cloth, wallpaper and some other materials of soft temperament, the whole space displays some other life interests. The space layout is simple but clear. Simple and orderly lines add some three-dimensional feel for the space. With the lights, the space appears to have more layers. The space boasts sedate and generous temperament. With delicate decorative details, the space seems to have more life interests.

Guest Room

Suite 1018

套房 1018

客厅融合了庄重与优雅的双重气质，使整个空间沉浸于典雅的贵族气息中。素雅的色调与利落的线条烘托出高贵的氛围和时尚的气质，并透过不同花纹的地毯、布艺来营造不同的空间氛围，让空间更具张力，给人以开放宽容的非凡气度。

The living room integrates dual temperaments of solemness and elegance, making the whole space get submerged in elegant aristocratic sphere. The elegant color tone and brisk lines set off noble atmosphere and fashionable temperament. The designer also creates different space atmosphere through carpets and cloth of different fabrics, entrusting the space with more tension and creating generous and spectacular temperament.

LAN TIAN HOTEL

Presidential Suite

总统套房

11F Presidential Suite

11F 总统套房

总统套房想传递一种低调奢华的生活方式，空间有着欧式生活的贵气与精致，但无金碧辉煌的浮华。每个空间都有自己独特的一面，新古典家具为空间增添了优雅华丽的气氛，与简明大气的空间格局搭配，更添端庄的韵味。客厅追求大方与开阔的空间感，同时表现奢华，私密空间则是温馨舒适。铜金色、银色、实木色等色彩与精美的雕花艺术搭配在一起，奢华而不流于俗气，表现真正对高端生活的追求。

This presidential suite aims to send out some low-key and luxurious lifestyle. This space has the nobility and delicacy of European life, but with no splendor vanity. Every side has its spectacular side. Neo-Classical furniture adds some elegant and resplendent atmosphere. Combined with concise and grand space layout, some solemn charms are created. The living room aims to produce generous and broad space feelings. While with the luxurious feelings, the private space appears warm and comfortable. Colors such as copper gold, silver and solid wood integrate perfectly with exquisite artistic carving. The design is vulgar but with no vulgar aspect, which displays the designer's pursuit of high-level life.

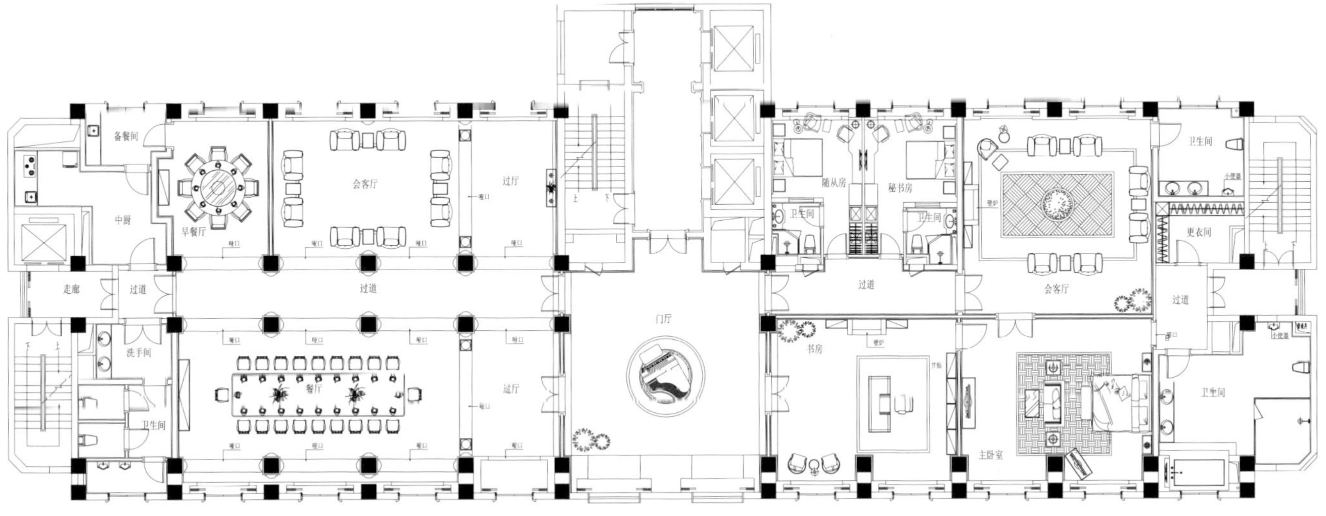

11F 总统套平面布置图

11F 总统套房 / 会客区

会客厅的整体空间简洁大方，流畅的线条与富有质感的材质组合，让空间立体感倍增。简洁的线条和精致材质的选用，也让整个环境充满着大气、低调的奢华感。家具的选择和饰品的搭配都恰到好处，家具精美而典雅，将空间的气质和内涵表达到位，点到为止的装饰让空间显得干净利索。

The concise and grand reception room boasts the combination of fluent lines and materials of fine texture, creating three-dimensional sense for the space. The application of concise lines and delicate materials makes the whole environment filled with gorgeous and low-profile luxury feel. The selection of furniture and collocation of ornaments are all proper. The delicate and elegant furniture displays the temperament and connotations of the space exactly, while the appropriate decorations make the space appear clear and tidy.

11F 总统套房 / 贵宾接待厅

贵宾接待厅将空间的简约之美与独树一帜的设计融为一体，营造出典雅的氛围。在确定主色调时，采用平和的淡色调进行整体装饰，透过墙面的石膏纹样展现出低调的华丽感。材料方面采用融入优雅的新古典家具，跳脱出雍容华贵的单调和千篇一律，精彩于此空间中表现得恰到好处。

The reception hall integrates the concise beauty of the space with the unique design, displaying elegant atmosphere. While selecting tone colors, the designer applies soothing light color in the whole decoration and presents low-key magnificence through the plaster pattern on the wall. As for materials, there are Neo-Classical furniture of elegance with no monotonous nobility or repetition. What a brilliant space.

11F 总统套房 / 餐厅

环境清幽。餐厅周围花团锦簇，成为酒店中央一片安静的绿洲。餐厅是欧式风格的完美表现，方形的长桌可满足多人聚餐的需要，背景墙采用油画装饰，散发着浓厚的欧式文化气息。餐厅以大理石结合水晶吊灯装饰，餐椅采用丝绒材质，一旁的立柱有着金色的古典纹样点缀其上。置身餐厅，感觉仿佛是在欧式古堡中就餐。

This is a serene location. The dining hall is surrounded by blossoming flowers, being a quiet oasis in the center of the hotel.
The dining hall is the perfect presentation of European style. The square long table can meet with the needs to accommodate many people to enjoy dinner together. The background wall is decorated with paintings, emitting intensive European cultural atmosphere. The dining hall is decorated with marble and crystal chandeliers. The dining chairs apply silk, while the pillars beside is designed with gold classical patterns. Being inside the dining hall, one would feel like enjoying meals inside an old European castle.

11F 总统套房 / 餐厅

餐厅中配备瑰丽的摆设，流露高雅感觉。吊灯，熠熠生辉，美妙绝伦。

餐桌正对着落地窗，在享受美食的同时还可以欣赏美景。

The dining hall is equipped with magnificent objects, presenting elegant sensations. The chandeliers are dazzling and superb.

The French window is opposite the dining table. Thus one can enjoy fine views while tasting gourmet.

Presidential Suite

11F 总统套房 / 书房

设计师运用恬淡的色彩加强设计感，同时以木质的硬包搭配呈现墙面的立体效果，将原本简约的空间表现得更加灵活多姿，大气中渗透着点点奢华，带给客人极致的生活及精神享受。

The designer applies peaceful colors to strengthen the design feel, while presenting the three-dimensional effect of wall through wood hard roll, making the original concise space appear more colorful and vivid. The magnificence is dotted with luxury, creating top-level life and spiritual enjoyments for the guests.

11F 总统套房 / 卧室

卧室设计舒适、通风，可以在这里享受到宁静舒适的环境，独立的设计使人感觉远离喧嚣闹市。卧室的主色调以白色与咖色为主，间以欧式花纹壁纸装饰，在温馨优雅中添上华贵的一笔，让其与整体空间氛围协调一致。

The cozy bedroom has fine ventilation. Inside this tranquil and comfortable environment, the design allows people to enjoy some private life far away from the busy urban life. The tone colors of the bedroom focuses on white and coffee colors.

11F 总统套房 / 主卫

主卫展现了现代欧式风格的简约、时尚、宁静的空间特点，在设计中，运用白色大理石作为装饰，让空间显得更加宁静和优雅。主卫配备超大观景浴缸及桑拿房，在符合精致主题的同时，带来奢华的生活体验。

This master washroom represents the concise, fashionable, and quiet space features of modern European style. For the design, the designer applies white marble as the decoration, making the space appear tranquil and elegant. The washroom is equipped with large observation bathtub and sauna room. While meeting with the exquisite theme, it also creates some luxurious life experiences.

12F Presidential Suite

12F 总统套房

总统套房的整体设计是奢华主义与现代浪漫主义的完美结合，流线形与直线相互搭配，现代奢华的家具精美华丽，明亮的色彩相互交错，个性的壁纸互相呼应。此情此景，充斥着浪漫的诱惑力与华贵的感染力，体现出王者风范，将华丽贵气的表达形成一套符号系统深植人心，将物体还原到最浪漫、纯粹的状态。

The whole design of the presidential suite is the perfect combination of luxury and modern romanticism. Streamline collocates with straight line, the modern luxurious furniture is exquisite and magnificent, bright colors interlace with each other and unique wallpaper corresponds with each other. Within this situation and this view, the space is overflowing with romantic seduction and noble attractions, demonstrating royal mode. The grand and superior representations form a set of symbol systems which become rooted in people's hearts, everything getting back to the most romantic and purest status.

12F 总统套房 / 进厅

进厅的设计有着强烈的进深感，繁复的纹样装饰配上华丽的水晶灯装饰，彰显出华贵的气息。进厅展台上鲜艳、娇嫩的花束是空间的一抹亮色，成为了空间视觉的焦点。空间的色彩随着空间的进深形成渐变的效果，使空间犹如音符般富有高低轻重的节奏感。

The design of the entrance hall has intensive depth. The complicated pattern is collocated with magnificent crystal lights decorations, presenting noble atmosphere. The fresh and bright bouquet on the counter is the highlight of the space, being the visual focus of the space. The space colors possess some gradual change with the continuation of the space, making the space acquire some musical notes like rhythm with variations.

Presidential Suite

12F 总统套房 / 贵宾接待厅

设计师将空间的平面、立面范畴都进行了归纳改造，划分出流畅优美的动线及完整、和谐的功能分区。大胆巧妙地运用天然石材及镜面光影，丰富了空间的视觉效果。舒畅而富有变化的空间，通过对欧式古典元素的融合创新，体现出独特的文化氛围，超然于奢华之外。

The designer redesigns the surface and elevation of the space, creating fluent moving lines and integral and harmonious functional zones. The designer courageously and ingeniously makes use of natural stone and mirror light and shadow to enrich the visual effects of the space. The smooth space rich in variations represents peculiar artistic atmosphere with integration and innovation of classical European elements, reaching far beyond luxury.

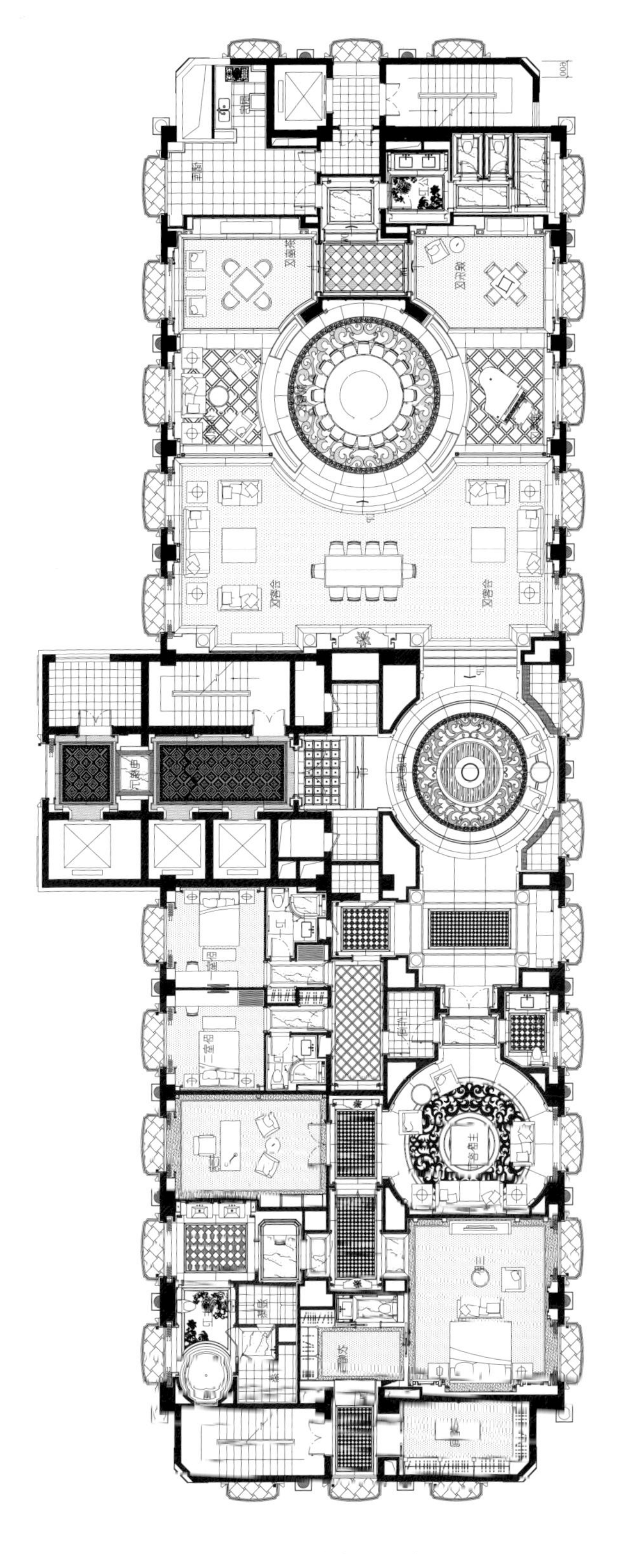

12F 总统套平面布置图

Presidential Suite

12F 总统套房 / 细节

空间的设计诠释出新古典的柔和温存，奢华而又婉约，层层递进的视觉环境容易与客人产生共鸣，激发内心的感动。 每一个细节都是对品味生活的理解，生活空间需要的不光是绚丽的外表，更注重的有品位的生活方式。

The space design interprets Neo-Classical softness and warmth, being luxurious and graceful. The progressive visual environment can easily produce resonance with the guests, provoking inner impressions. Every detail is the appreciation in tasting life. What life space needs is no flamboyant exterior, but lifestyle of fine taste.

12F 总统套房 / 餐厅

圆形、弧形与直线在空间演奏着幽雅的乐章，流畅地完成了空间的转换。背景墙的设计温馨大气，与整体空间浑然一体，却又显得立体而又生动。时尚的餐厅空间搭配实木的餐台，种种设计理念为客人带来舒适体验。木质的不同形式的桌子，分布在不同的区域，保证餐厅空间得到最大限度的利用。

Round, arch and straight forms produce elegant musical chapters inside the space, smoothly completing the space transitions. The design of the background wall is warm and gorgeous, making the whole space like a whole, but appearing three-dimensional and vivid. The fashionable dining space is collocated with solid wood dining counter. Various kinds of design concepts aim to bring comfortable experiences for guests. Wood tables of different forms are displayed in different zones, guaranteeing that the dining space can be made full use of.

12F 总统套房 / 书房

为了保证室内的采光度，并有效地利用室外的风景，书房以高大的落地窗为室内提供光照，可以让光线畅通无阻地照进空间。室内的装饰优雅而沉静，适合客人读书、办公、会客的需求。

In order to maintain the interior lighting quality and make effective use of exterior sceneries, the study applies large French window to introduce illumination for the interior space, thus lights can spread inside the space with no obstructions. The interior decoration is elegant but sedate, quite suitable for guests' requirements in reading, work and meeting guests.

12F 总统套房 / 卧室

卧房的设计犹如一段优美的音乐，它散发出的力与美就像音乐的节奏和韵律，冲击着人的心灵，带给人美的享受。设计师用现代的手法和材质还原古典气质，于是，整个空间便兼具了古典和现代的双重审美效果 在色彩和配饰的搭配上，以凸显典雅、华贵和沉稳为主，不造作、不刻意，将空间演绎得流畅而利索，让身处其中的人感受生活的美好和高雅的格调。

The design of the bedroom is like a piece of elegant music, sending out power and beauty which is like the rhythm and rhyme of music, impacting people's heart and creating aesthetic enjoyment for people. The designer uses modern approach and materials to restore classical temperament. Thus, the whole space possesses dual aesthetic effects of classical and modern sphere. As for the collocation of colors and ornaments, the designer focuses on elegance, nobility and profoundness. The natural design presents the space in a fluent and tidy way, making people inside feel the nicety of like and the elegant tone.

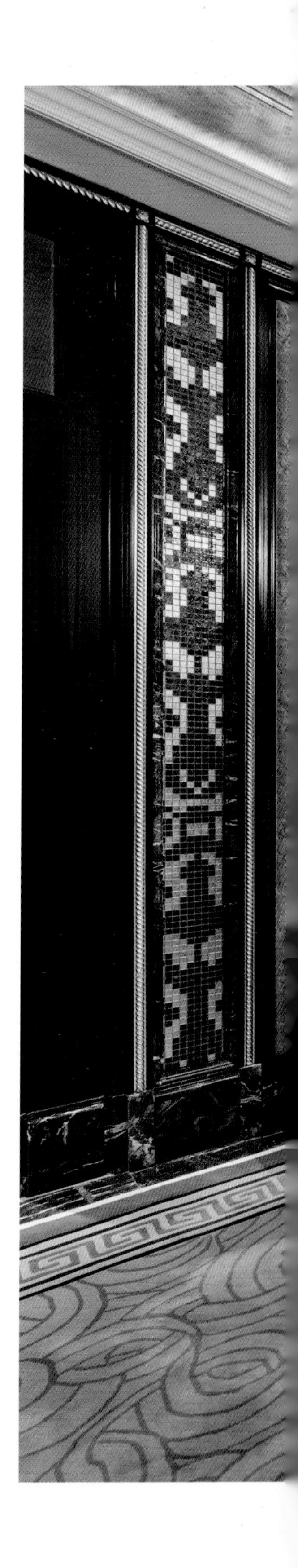

12F 总统套房 / 主卫

主卫的装修风格在材质及纹样的选取上，与其他空间保持了高度的和谐。其装修风格唯美、大气，是理想的时尚休闲场所。卫浴设施都是考虑了人使用时的舒适度精心设计的，弧形的观景窗可以让客人最大范围地欣赏到窗外的美景，在这里，客人可以放松身心地享受一切。

As for the selection of materials and patterns, the decoration style maintains harmony with other spaces. With this nice and grand decorative style, this is a perfect leisure location for fashion and entertainment. The bathroom facilities are carefully designed which takes people's comfort while use into considerations. The arch observation window allows the guests to have a broad view of the scenerics outside. Here, the guests can get fully relaxed and enjoy everything in the hotel.

SONY

图书在版编目(CIP)数据

走进蓝天：蓝天酒店/ 蓝天酒店编. 一武汉 ：华中科技大学出版社，2013.9
ISBN 978-7-5609-9283-9
Ⅰ. ①走… Ⅱ. ①蓝… Ⅲ. ①饭店－室内装饰设计－济南市－图集 Ⅳ. ①TU247.4-64

中国版本图书馆CIP数据核字(2013)第182305号

走进蓝天·蓝天酒店 蓝天酒店 编

出版发行：华中科技大学出版社（中国·武汉）
地　　址：武汉市武昌珞喻路1037号（邮编：430074）
出 版 人：阮海洪

责任编辑：曾　晟　　责任监印：秦　英
责任校对：赵爱华　　装帧设计：张　艳

印　　刷：北京利丰雅高长城印刷有限责任公司
开　　本：787 mm×1092 mm　1/12
印　　张：22
字　　数：132千字
版　　次：2013年9月第1版第1次印刷
定　　价：498.00元 （USD 89.99）

投稿热线：(010)64155588-8000 hzjztg@163.com
本书若有印装质量问题，请向出版社营销中心调换
全国免费服务热线：400-6679-118　竭诚为您服务